NEXT-GENERATION MILITARY TECHNOLOGY

SPACE DEFENSE SYSTEMS

IN THE MILITARY

by Cynthia Kennedy Henzel

BrightPoint Press

San Diego, CA

BrightPoint Press

an imprint of ReferencePoint Press, Inc.
Printed in the United States

For more information, contact:
BrightPoint Press
PO Box 27779
San Diego, CA 92198
www.BrightPointPress.com

LIBRARY OF CONGRESS CATALOGING-IN-PUBLICATION DATA

Name: Henzel, Cynthia Kennedy, author.
Title: Space defense systems in the military / by Cynthia Kennedy Henzel.
Description: San Diego, CA: ReferencePoint Press, 2027 | Series: Next-generation military technology | Includes bibliographical references and index. | Audience: Grades 7–9
Identifiers: ISBN 9781678214265 (hardcover) | ISBN 9781678214272 (eBook)
The complete Library of Congress record is available at www.loc.gov.

CONTENTS

AT A GLANCE

- Many satellite systems are used to defend the United States.
- The US Space Force is the newest US military branch. It is responsible for using space technology to defend the United States.
- Space Command is made up of members of different military branches. They carry out military operations outside Earth's atmosphere.
- The Space Based Infrared System (SBIRS) detects infrared energy. People feel infrared energy as heat.
- SBIRS helps track missile launches and nuclear tests. It spots infrared energy from these events.
- The Global Positioning System (GPS) uses satellites to find and track objects on Earth.

- Space junk is a threat to satellites in space.
- Spy satellites in space can take pictures of something smaller than a person.
- Spy satellites can intercept military or government communications.
- New anti-satellite weapons threaten the peaceful use of space.

INCOMING ATTACK!

On January 8, 2020, ballistic missiles were launched from western Iran. US satellites spotted them right away. Satellites are objects placed in orbit around Earth. They collect information or communicate. The satellites sent an alert to Buckley Air Force Base in Colorado. The missiles appeared on screens at the operations center. **Analysts** figured out

In 2023, thirty-one countries had ballistic missiles, while nine nations were believed to have nuclear weapons.

where the missiles came from. They also found the missiles' target.

The missiles were headed for a military base in Iraq. The base is called Ain al-Asad.

American troops at Ain al-Asad base in Iraq were targeted in many drone and rocket attacks between 2023 and 2025.

Hundreds of Americans lived there. Analysts in Colorado sent a warning to the base. They used satellite communications. The message was clear. Missiles would hit in 15 minutes. The Americans took shelter. They hid in underground bunkers.

The first missiles hit at 1:34 a.m. The attack lasted 2 hours. Buildings became rubble. The blast injured 109 Americans. Some had serious brain injuries. They were caused by the blast waves. But no one was killed.

EYES IN THE SKY

The United States has a space defense system. It has many groups of satellites. They are controlled from the ground. One is the Space Based Infrared System (SBIRS).

It warns of missile launches. The SBIRS saved the Americans at Ain al-Asad base.

Two satellite groups collect data on weather. Others track everything that circles Earth. This includes space junk. Space junk includes any unused objects floating in space. It can hit and damage spacecraft.

The Global Positioning System (GPS) helps people find their way. The military uses GPS to find **assets**. It also helps locate enemies. Other systems help people communicate. Lieutenant General Steven L. Basham was a leader in the Air Force. He explains, "Space operations are woven into the fabric of modern military operations."[1]

The GPS system used by military personnel has Precise Positioning Service. It provides locations within 1 inch (3 cm) compared to 10 to 16 feet (3 to 5 m) for standard systems.

CHAPTER ONE

THE SPACE FORCE

The US Space Force is a key part of the space defense system. Its members are called Guardians. The US Space Force helps defend the country. To do this, the Space Force uses space technology.

The Space Force has a base in Florida. It has two more in California and three in Colorado. There are about 9,400 active-duty Guardians. The Space Force has combat groups. Each has different

Space Force was formed in December 2019. It is the newest branch of the military.

UNITED STATES SPACE FORCE
MMXIX

types of experts. Some are experts in intelligence. Others make sure US missiles find their targets.

The Guardians support missions. They train to stop enemy threats. Enemies could use lasers to blind US satellites or block communications. The Space Force provides commanders with tools to do their jobs. These include satellites to detect the movement of troops.

Guardians begin with basic military training. They learn how to be Guardians. This includes everything from saluting to marching. Guardians exercise to become physically fit. They learn discipline. They also learn how to work together. Guardians then train to support space defense. They learn to operate satellite systems. The Space

Force needs people with strong engineering and technology skills.

UNITED STATES SPACE COMMAND

The US Space Command is another part of the space defense system. Space Command was founded in 1985. It is not part of a single military branch. Instead, Space Command directs forces from

From High School to Space

Jobs for high school graduates in the Space Force include targeting analysts. These people help find targets. They also detect missile threats. High school graduates can become space systems operators. These Guardians track satellites. Guardians who do this work can earn college credits through the Community College of the Air Force.

all military branches. They carry out military operations in space. This means anything 62 miles (100 km) from the Earth and beyond.

General Stephen N. Whiting heads the US Space Command. He explains how Space Command works. He says, "I am

Space Command officers toured a launch facility at Cape Canaveral Space Force Station in Florida in 2021.

a member of the US Space Force, but I have Army **personnel**, Navy personnel, Air Force, Marine Corps personnel, even some international partners in our headquarters."[2]

Space Command is smaller than the Space Force. Space Command is located at Peterson Space Force Base in Colorado. But in 2025, President Donald Trump made an announcement. Space Command was moving to the Redstone Arsenal in Alabama. This is the US Army's center for missile and rocket programs.

A RACE TO SPACE

In 2025, almost 13,000 satellites orbited Earth. The United States had launched more than 8,500 of them. This is more than any other country. Some of these belong to the National Aeronautics and Space Administration (NASA). The military has several groups of satellites. Others belong to companies such as SpaceX. Its Starlink project has more than 8,000 satellites. They provide internet service around the world.

Space defense is key to military operations. It provides data about nuclear testing. It tracks enemy troop movements. Satellites help nations communicate quickly. But space defense goes beyond the military. Satellites help everyone communicate. They help **monitor** and gather weather data. This helps protect people from floods or storms.

Satellites are used for television broadcasts. They help cars, boats, and planes **navigate**. Banks and cell phone

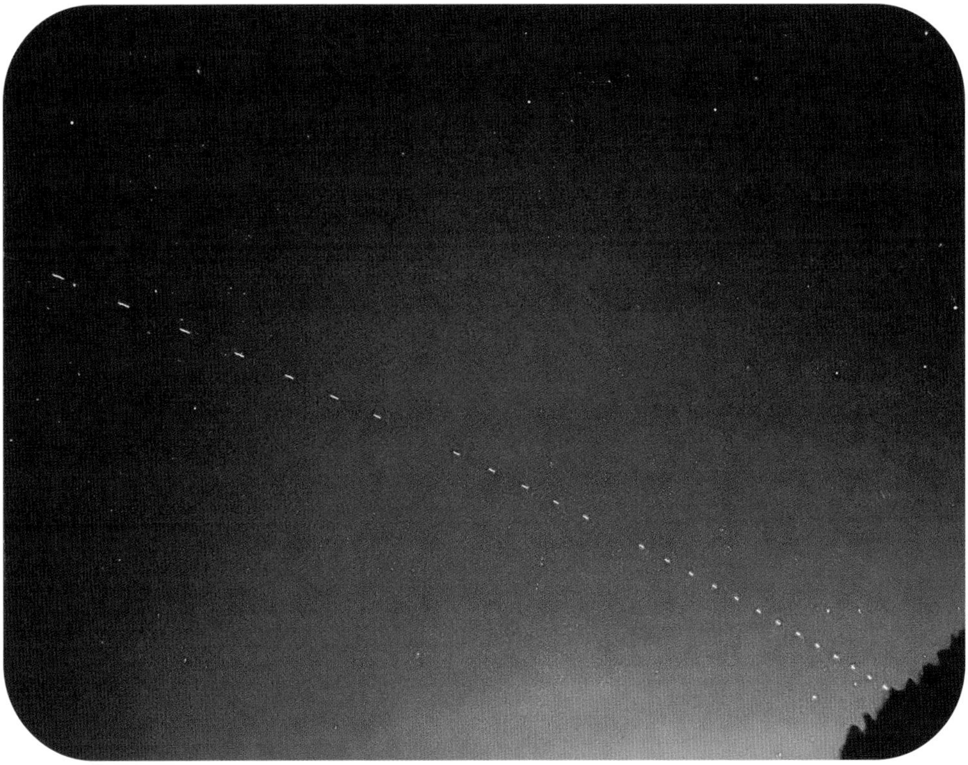

On a clear night, people on Earth can see a trail of Starlink satellites when they are first deployed.

networks communicate through satellites, too. People depend on satellite systems every day. Space defense keeps enemies from attacking these systems.

Many other countries have satellites. In 2025, Russia had more than 1,500. China had over 900. These countries are quickly adding more satellites. They hope to challenge the power of the United States in space.

Most satellites are used for peaceful purposes. These include communication and research satellites. But some countries threaten other countries' satellites. Russia, China, the United States, and India have

ASAT weapons can be launched from stationary sites on Earth or from aircraft such as an F-15 Eagle.

tested anti-satellite weapons (ASATs). In 2015, Russia and China began developing ground missiles that could destroy satellites.

China launched a new kind of satellite in 2016. It can fly close to other satellites. It might target US satellites. According to the Defense Intelligence Agency, "[Russia and China] view space as important to modern warfare."[3] During wars, other countries might target satellites to harm the United States. US space defense systems protect the nation's space assets.

CHAPTER TWO

SPACE BASED INFRARED SYSTEM

SBIRS senses infrared energy. Infrared energy is a part of the light spectrum. People see visible light. They do not see infrared light. But they feel it as heat. The SBIRS senses heat from any source. Missiles create heat. Forest fires cause heat. Even people give off heat. Analysts are trained to identify the source of infrared energy.

SBIRS Geostationary, or GEO, satellites are launched from Cape Canaveral Air Force Station in Florida. They hold the same position 22,400 miles (36,000 km) above the Earth's equator while traveling through space.

The SBIRS scans the entire Earth. It takes pictures every 10 seconds. If it finds suspicious areas of heat, it notifies analysts on the ground. Operators can then decide what to do. SBIRS is also able to scan smaller areas. This lets SBIRS take close-up images of missile launches.

The SBIRS is made up of several parts. There are older Defense Support Program (DSP) satellites. There are newer Geostationary Earth Orbit (GEO) satellites. Six GEOs orbit Earth above the equator. Each satellite stays over the same place as the Earth turns. Other satellites in the SBIRS carry High Earth Orbit (HEO) sensors. They cover Earth's poles.

Ground stations are another part of the SBIRS. These collect data from

Ground stations receive data from and send data to satellites as they pass overhead.

the satellites. Analysts then use the data to understand what they are seeing.

NEW GEOS

The GEO satellites were put into orbit between 2011 and 2022. Each one is about seven times the size of a small bus. This does not include its solar panels. The panels power the satellite. GEOs resist damage better than the DPS satellites. They are made to last longer.

The GEOs also sense a broader range of infrared energy than DSP satellites.

Infrared sensors identify items that give off heat, such as ballistic missiles, humans, explosions, or fire.

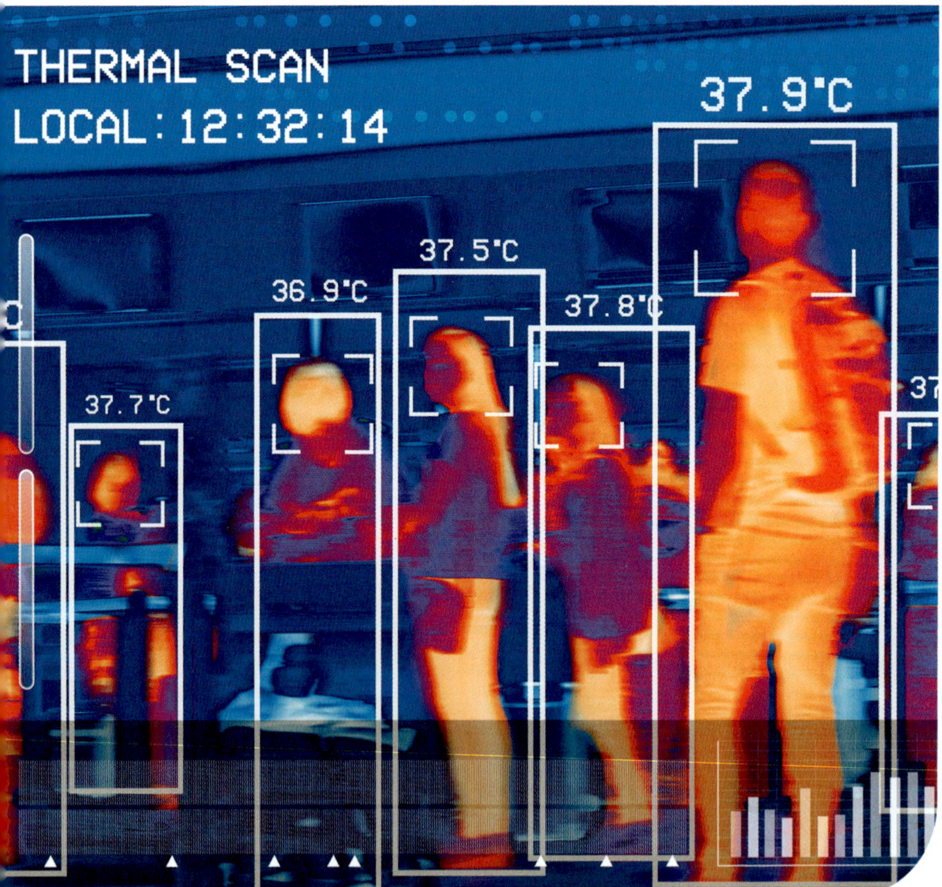

This helps them identify different types of missiles. This is because missiles produce heat while traveling through the atmosphere, or air. Ballistic missiles create a burst of heat in the air. Then they travel above the atmosphere. In space, they produce little heat. Warheads on rocket boosters travel at very high speeds through the atmosphere. They produce a lot of heat. Cruise missiles fly at lower speeds. So, they produce less heat.

Analysts use infrared sensors to identify the type of missile. They can tell how fast it is going and in which direction. They can predict the likely landing point. The system was used in October 2024. Iran launched 200 ballistic missiles at Israel. It was the largest ballistic missile attack in history.

The United States and its allies destroyed some of the incoming missiles. Only one person was killed in the attack.

GEOs have a sensor called a step-starer. It can be pointed at an exact location. The step-starer provides more data about a smaller location. The GEOs can process some data on-board. Then they can send detailed information quickly to operators on the ground.

SBIRS detects space launches. This means the United States knows what is sent into space. Plus, it knows who sent the object. The system also detects nuclear blasts. The United States can tell who is testing nuclear weapons. This data helps enforce agreements between countries. It also helps identify nuclear threats.

SBIRS can also detect non-military events. It detects forest fires. Different types of explosions are recorded. In 2014, SBIRS detected an explosion. Malaysia Airlines Flight 17 had been hit with a missile. Analysts identified the missile as a

The SBIRS GEO-2 was launched into space on March 19, 2013, from Cape Canaveral Air Force Base in Florida.

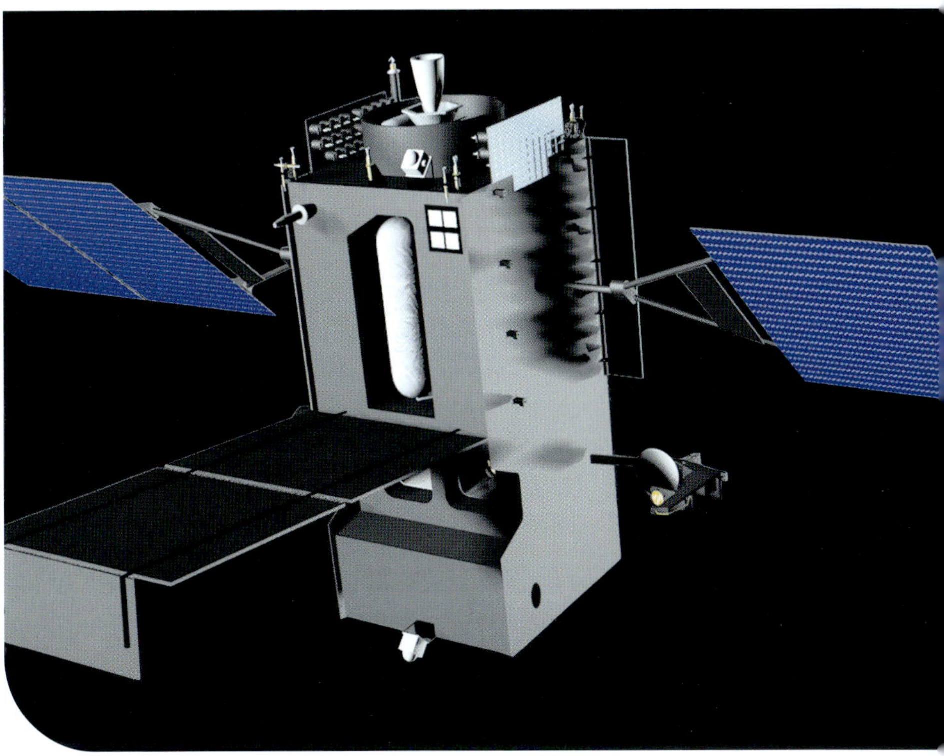

Russian one. They traced it back to where it had been fired.

LOOKING AHEAD

Jeff Schrader is a vice president at Lockheed Martin. The company builds many satellites. He says, “America’s best defense is always going to start with having the earliest possible warning of threats—and that warning comes from what we see in space.”[4] SBIRS can detect missile launches faster than any other system. But the military must stay ahead in technology.

The Space Force is ready to launch the Next-Generation Overhead Persistent Infrared (Next-Gen OPIR) satellites. This will replace the SBIRS. The planned launch for the first satellite was in 2026.

Next-Gen OPIR satellites detect missiles that are dimmer and have shorter burn times when launched. This is helpful with hypersonic missiles. These are missiles that travel at more than five times the speed of sound. The satellites are built to withstand new threats from other countries.

Faster and Faster

The speed of sound is about 760 miles per hour (1,223 kmh) at sea level. This is called Mach 1. Most aircraft travel slower than Mach 1. Military aircraft often travel at supersonic speeds. This is between Mach 1 and Mach 5. Hypersonic speeds are faster than Mach 5. Most metals melt at these speeds. Aircraft must have special protection from heat.

CHAPTER THREE

GLOBAL POSITIONING SYSTEM

The GPS is a group of thirty-one satellites. It gives exact locations anywhere on Earth. The system is owned by the United States. But it can be freely used by anyone in the world.

A GPS satellite is about the size of a small bus. Each satellite includes a clock. The clock is accurate to one-millionth of a second. The satellites orbit the Earth every 12 hours. They constantly beam

Lockheed Martin developed the thirty-one GPS satellites currently orbiting the Earth.

radio waves back to Earth. A GPS device on Earth receives the radio waves. It then calculates how long it took the waves

Rockets such as SpaceX's *Falcon 9* carry GPS III SV-08 satellites into orbit. Adding new satellites to the GPS constellation improves the accuracy of the data.

to arrive. It measures waves from three satellites to calculate a location. Waves from a fourth satellite can provide elevation.

The first GPS had 24 satellites. It was completed in 1993. The United States worried other countries could use the system to target US ships or planes. It allowed people outside the US military to use GPS. So it added errors to the measurements. Only the military could use GPS to find exact locations. In 2000, these errors were turned off.

GPS IS EVERYWHERE

Today, GPS is part of everyday life. GPS receivers are in smartphones and cars. Airports use the system to keep planes safe. Farmers use GPS, too. It helps them guide tractors and monitor crops. Construction projects depend on GPS for locating sites. The system also remains key to national defense.

The military still controls GPS signals. The newer GPS III satellites are accurate to 3.3 to 9.8 feet (1 to 3 m) for civilian users. Accuracy is affected by buildings or trees. These can block the signal.

GPS accuracy can be improved with a ground-based station. These stations give out a GPS signal. Real-Time Kinematic (RTK) GPS is used for mapping. RTK GPS devices receive signals through the internet. The signal provides data that corrects the regular GPS signal. RTK GPS is much more accurate than GPS III. RTK GPS is accurate to within 0.4 inches (1 cm).

A surveyor may use a drone equipped with RTK GPS to more accurately map an area.

GPS IIIF

All military assets have GPS. It helps land vehicles get through remote areas. GPS is used to pilot drones. Ships and submarines use GPS. Soldiers wear GPS devices.

HOW GPS WORKS

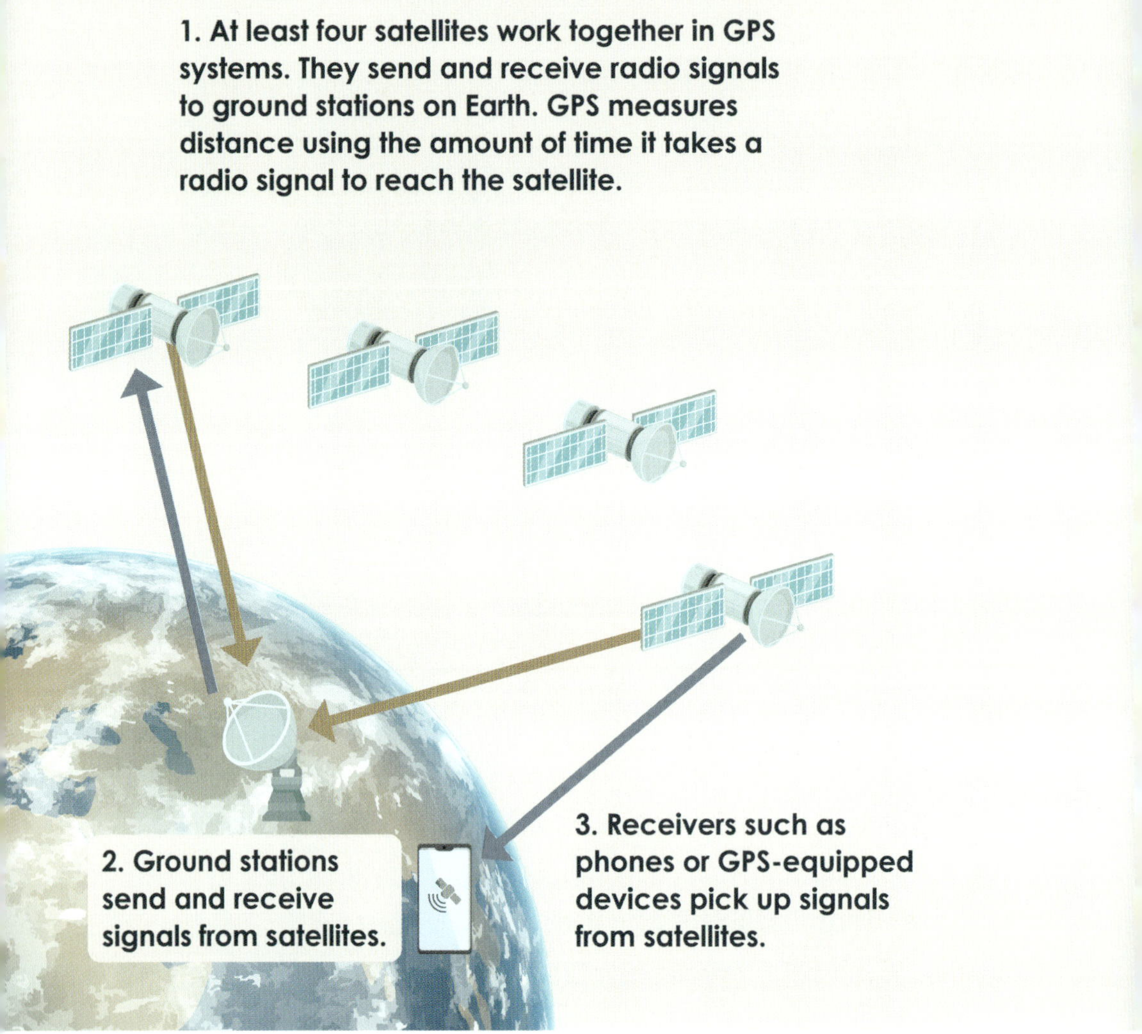

GPS is a three-part system. It includes satellites, ground stations, and receivers.

These devices provide location coordinates for the vehicles, drones, ships, and soldiers. This can help fellow soldiers find them.

GPS IIIF is the newest US system. The first satellites launched in 2018. By 2025, eight were in orbit. But only the military and NASA could use the new system.

The new GPS IIIF satellites are more accurate than the GPS III. They can be upgraded while in orbit. A special feature on GPS IIIF is an improved M-code signal. This is an **encrypted** signal. It resists **jamming** and tampering. The new satellites can also detect distress calls. They then report the calls to police or fire officials.

GPS IIIF has a system to detect nuclear blasts. It has sensors for X-rays, electromagnetic pulses, and light. This helps it identify exactly where nuclear activity occurs. GPS IIIF signals can also be detected from the Moon. This will someday help travelers navigate on the Moon's surface.

BEYOND GPS

The US GPS system is controlled by Space Delta 8. This is a Space Force unit focused on communications. It is at Schriever Space Force Base in Colorado. Other systems

Space Force Fitness

A test program launched in 2023 provides Guardians with smartwatches. They can now wear these devices instead of passing physical fitness tests. The military uses data gathered by the watch to track the Guardian's fitness. Previously, military members could not wear smartwatches. This was because enemies could track them using GPS. But Guardians can now turn off the GPS if needed.

Delta 8 SOPS Guardians maintain the Mobile Constellation Control System at Schriever Space Force Base in Colorado.

are also in use. Russia, China, and the European Union have their own navigation systems. China's system has 56 satellites.

This gives it greater accuracy than GPS in many parts of the world.

Some military experts are worried about GPS navigation. They think the United States depends on it too much. Robert Hoffman is an expert in military space operations. He wants the United States to stop updating the old GPS system. He thinks the military should ask scientists to invent a better system. He says, “In any future conflict with [an equal enemy], the US military can . . . expect to operate in a GPS-denied environment.”[5] This means the military may not be able to depend on GPS to work.

Marines use lightweight GPS-equipped devices to measure and relay accurate location coordinates for targets in the field.

CHAPTER FOUR

INTELLIGENCE AND PROTECTING SPACE ASSETS

The United States uses some satellites to spy on other countries. These spy satellites provide pictures of activity on the ground. The Keyhole-11 group of satellites can identify objects that are the size of a car. The newer Keyhole-12 can see something 5 inches (12.7 cm) long. Experts believe even better satellites exist. But the government keeps them a secret.

A Delta IV rocket launched on December 10, 2020, with a NROL-44 spy satellite on board.

These spy satellites are powerful tools. But they do not work like in movies. They cannot quickly zoom in on a particular spot. They cannot take video as they whiz around Earth. Clouds and darkness may limit what they can see as well.

Another type of spy satellite listens to radio signals. The NROL-44 is one of the most powerful spy satellites in the world. It has huge antennas. Some may be 328 feet

First Spy Satellites

The Corona satellites were the first US spy satellites. They were used between 1959 and 1972. Coronas had cameras that took pictures. The images appeared on plastic strips called film. Then the film container was ejected from the satellite. Airplanes with special hooks caught the containers as they fell to the Earth.

(100 m) wide. NROL-44 can **intercept** military or government communications.

SORTING THROUGH DATA

Space Delta 7 is a unit of the US Space Force. Analysts there use data from spy satellites to create intelligence. Intelligence is information gathered to protect a country from threats. Leaders use this intelligence to make decisions.

The United States has a lot of satellite data. It is hard for the analysts to keep up. Artificial intelligence (AI) helps examine data. AI can do this faster than humans. AI can find a specific object in an image, such as a missile launcher. It then puts a box around the object. This helps the analyst identify a threat more quickly.

AI can count objects. It can organize them into groups. These might include different types of satellites. Or AI could identify different sizes of space junk.

There are about 23,000 pieces of space junk orbiting Earth that are bigger than 4 inches (10 cm). Items include old spacecraft, rocket boosters, an astronaut's glove, and even a toothbrush.

AI models notice when something unexpected happens. This could be a satellite exploding or changing its path.

AI models are being trained to predict what will happen next. Then they will find more information to see if the prediction is correct. Patrick Biltgen is an AI expert. He discussed how AI could help study all this data. He says, "The system knows what missing piece of information is needed, and it goes and gets it. That would be a breakthrough in intelligence."[6]

SPACE TRAFFIC

More satellites go up each year. Plus, space junk circles the globe. When one satellite broke up in 2024, it created 700 pieces of space junk. Space junk is traveling fast.

This means even a tiny piece of space junk could cause damage if it hits a satellite. The US Space Surveillance Network (SSN) tracks more than 47,000 objects that circle Earth. This is a difficult job. It is becoming more difficult.

Space trackers must also pay attention to what each object is doing. In 2022, China used a spacecraft to move one of its satellites to a new orbit. This ability to grab and move satellites is a danger to US assets. Trackers now look for strange movements from enemy satellites.

New software is helping the military track objects in space. It is called iSpace. It sends an alert if an object acts strangely. This software has the ability to track more than 300,000 space objects.

Another tool is Space Fence. It is a ground-based radar system. Space Fence can detect an object the size of a marble. Space Fence tracked the space junk from an Indian anti-satellite test in 2019.

At the Optical Observatory for Pico do Areeiro in Madeira, Portugal, scientists look for and monitor space junk. The site is part of the European Space Surveillance and Tracking program.

It predicted where the pieces of the satellite would be as they circled Earth.

WEAPONS IN SPACE

The United States tested the first anti-satellite weapon (ASAT) in 1959. These weapons could destroy an enemy satellite. They were fired from a plane or another satellite. To test ASATs, countries blow up useless satellites. The tests cause a lot of space junk. In 2022, the United States announced it would no longer test ASATs. Most countries of the United Nations supported the ban. But Russia, China, and India did not.

The United States worried that an ASAT could harm their satellite systems. The country decided to move to a new type

Space Force first lieutenant Tia Scoggan monitored the skies for threats in Earth's orbit.

of satellite defense. They would not build one big, costly satellite to do a job. Instead, they would use many small satellites. If an enemy destroys one small satellite, others can still do the job. This makes the use of ASATs less likely. It is not worth the cost of a missile to take out a small satellite.

The United States launches new satellites frequently. Each group of new satellites is better than the last. Derek Tournear is the

Private companies such as SpaceX have significantly increased the number of satellites launched each year. In late 2024, 11,539 satellites were orbiting Earth, compared to 3,371 in 2020.

director of the Space Development Agency. It develops and launches the new satellites. "**Proliferation** is our biggest defense," Tournear says.[7]

On May 16, 2024, Russia launched Cosmos 2576. This new satellite was in the same orbit as a US satellite. Russia placed Cosmos 2576 there to watch and track the US satellite. Russia said the satellite was there to inspect other satellites. But it might also be able to harm other satellites.

Other Russian satellites were suspect, too. One was Russian satellite Cosmos 2553. The United States thought it was part of a nuclear space weapon program. A nuclear weapon could destroy many satellites at once. This would defeat the idea of using many small satellites. Cosmos 2553

In late 2025, approximately 1,500 Russian satellites and 8,000 US satellites orbited the Earth. Many other countries also have satellites in space.

was launched in 2022. It was orbiting Earth just weeks before Russia invaded Ukraine. A nuclear space weapon could destroy the SpaceX Starlink system. Ukraine depends on Starlink for communication. This is because much of Ukraine's infrastructure has been destroyed in the war.

The United States has opposed weapons in space. But that may change. Other countries are developing new tools to take out satellites. In 2024, General Stephen N. Whiting called for the United States to put weapons in space. He said, "We need them to deter a space conflict and to be successful if we end up in such a fight."[8] Staying ahead in technology may help prevent a war in space.

GLOSSARY

analysts

people who use data to solve problems

assets

any resources controlled by the military, including personnel, weapons, vehicles, aircraft, and infrastructure

encrypted

turned into code

intercept

to secretly receive a communication signal

jamming

emitting a signal that blocks the GPS receiver from getting a signal from a satellite

monitor

to watch something over time

navigate

to use tools to find a way to a location

personnel

people who work for a company or the military

proliferation

a quick increase in the number of something

SOURCE NOTES

INTRODUCTION: INCOMING ATTACK!

1. Lt. Gen. Steven L. Basham, "The Critical Role of Space in Modern Warfare," *United States European Command*, February 9, 2024. www.eucom.mil.

CHAPTER ONE: THE SPACE FORCE

2. Quoted in Derek Lacey, "Space Command and Space Force Are Not the Same Thing," *AXIOS Huntsville*, September 16, 2025. www.axios.com.

3. Quoted in W. J. Hennigan, "America Really Does Have a Space Force," *Time*, January 23, 2020. https://time.com.

CHAPTER TWO: SPACE BASED INFRARED SYSTEM

4. Quoted in "First Next-Gen GEO-Based Missile Warning Satellite Successfully Completes Environmental Testing," *Lockheed Martin*, August 6, 2025. https://news.lockheedmartin.com.

CHAPTER THREE: GLOBAL POSITIONING SYSTEM

5. Robert Hoffman, "Lost on the Next Battlefield," *War Room*, January 21, 2022. https://warroom.armywarcollege.edu.

CHAPTER FOUR: INTELLIGENCE AND PROTECTING SPACE ASSETS

6. Quoted in "Artificial Intelligence for Space Missions," *Booz Allen*, n.d. www.boozallen.com.

7. Quoted in C. Todd Lopez, "Proliferation Remains Best Deterrence Against Threats to U.S. Space Access," *U.S. Department of War*, March 27, 2024. www.war.gov.

8. Quoted in Audrey Drecker, "The US Needs 'Weapons in Space,' SPACECOM Head Says," *Defense One*, April 8, 2025. www.defenseone.com.

FOR FURTHER RESEARCH

BOOKS

Michael H. Gorn and Davide Sivolella, *The Complete Book of Spacecraft: Rockets, Shuttles, Satellites, and Space Stations*. Motorbooks, 2025.

Cynthia Kennedy Henzel, *Cyber Warfare in the Military*. BrightPoint Press, 2027.

Chelsea Xie, *Life in the Delta Force*. BrightPoint Press, 2024.

INTERNET SOURCES

"GPS," *National Geographic*, 2025. https://education.nationalgeographic.org.

"What Are Infrared Waves?" *NASA*, n.d. https://science.nasa.gov.

"What Is a Satellite?" *NASA Science Space Place*, September 26, 2025. https://spaceplace.nasa.gov.

WEBSITES

GPS

www.gps.gov

GPS is an official government website for the Global Positioning System. It has information on what GPS can do and who runs it.

NASA

www.nasa.gov

The official website for the National Aeronautics and Space Administration (NASA) has information on all types of space subjects.

US Space Force

www.spaceforce.com

This is the official website for the US Space Force. It provides information on Space Force opportunities, training, and benefits.

INDEX

IMAGE CREDITS

Cover: © NicoElNino/Shutterstock Images
5: © zef art/Shutterstock Images
7: © maradon 333/Shutterstock Images
8: © Ayman Henna/AFP/Getty Images
11: © Milan Tomazin/Shutterstock Images
13: © NLM Photo/Shutterstock Images
16: Airman 1st Class Thomas Sjoberg/US Space Force/DVIDS
19: © Innovative Base Artistry/Shutterstock Images
20: Paul E. Reynolds/National Archives
23: US Space Force
25: © woodvillage/Shutterstock Images
26: © Have a nice day Photo/Shutterstock Images
29: US Air Force
33: US Air Force
34: © Manuel Mazzanti/NurPhoto/Getty Images
37: © MDV Edwards/Shutterstock Images
38 (satellites, phone): © Nsit/Shutterstock Images
38 (ground station): © Nsit/Shutterstock Images
38 (Earth): © Kolonko/Shutterstock Images
41: David Grim/US Space Force/DVIDS
43: Mathuel Browne/US Marine Corps/DVIDS
45: Jeff Spotts/US Space Force
48: © Frame Stock Footage/Shutterstock Images
51: © Sander van der Werf/Shutterstock Images
53: Tiana Williams/US Space Force/DVIDS
54: © aapsky/iStockphoto
56: © Mechanik/Shutterstock Images

ABOUT THE AUTHOR

Cynthia Kennedy Henzel has degrees in education and geography. She has written more than a hundred books for young people. These include biographies, fiction, and nonfiction on subjects such as geography, science, culture, and history. In the early 1990s, Henzel trained teachers to use GPS in classroom projects.